Hospital Fire Kills Four Patients
Southside Regional Medical Center
Petersburg, Virginia

Investigated by: J. Gordon Routley
Reade Bush

This is Report 080 of the Major Fires Investigation Project conducted by Varley-Campbell and Associates, Inc./TriData Corporation under contract EMW-94-C-4423 to the United States Fire Administration, Federal Emergency Management Agency.

FEMA

Department of Homeland Security
United States Fire Administration
National Fire Data Center

U.S. Fire Administration Fire Investigations Program

The U.S. Fire Administration develops reports on selected major fires throughout the country. The fires usually involve multiple deaths or a large loss of property. But the primary criterion for deciding to do a report is whether it will result in significant "lessons learned." In some cases these lessons bring to light new knowledge about fire--the effect of building construction or contents, human behavior in fire, etc. In other cases, the lessons are not new but are serious enough to highlight once again, with yet another fire tragedy report. In some cases, special reports are developed to discuss events, drills, or new technologies which are of interest to the fire service.

The reports are sent to fire magazines and are distributed at National and Regional fire meetings. The International Association of Fire Chiefs assists the USFA in disseminating the findings throughout the fire service. On a continuing basis the reports are available on request from the USFA; announcements of their availability are published widely in fire journals and newsletters.

This body of work provides detailed information on the nature of the fire problem for policymakers who must decide on allocations of resources between fire and other pressing problems, and within the fire service to improve codes and code enforcement, training, public fire education, building technology, and other related areas.

The Fire Administration, which has no regulatory authority, sends an experienced fire investigator into a community after a major incident only after having conferred with the local fire authorities to insure that the assistance and presence of the USFA would be supportive and would in no way interfere with any review of the incident they are themselves conducting. The intent is not to arrive during the event or even immediately after, but rather after the dust settles, so that a complete and objective review of all the important aspects of the incident can be made. Local authorities review the USFA's report while it is in draft. The USFA investigator or team is available to local authorities should they wish to request technical assistance for their own investigation.

This report and its recommendations were developed by USFA staff and by TriData Corporation, Arlington, Virginia, its staff and consultants, who are under contract to assist the Fire Administration in carrying out the Fire Reports Program.

The United States Fire Administration greatly appreciates the cooperation received from Fire Chief Jerry Wallace, Investigator Rufus L. Atkins, and Investigator David M. Jolly of the Petersburg, Virginia Bureau of Fire; and Robert A. Phelps, Director of Safety and Security at Southside Regional Medical Center.

For additional copies of this report write to the U.S. Fire Administration, 16825 South Seton Avenue, Emmitsburg, Maryland 21727. The report is available on the Administration's Web site at http://www.usfa.dhs.gov/

U.S. Fire Administration
Mission Statement

As an entity of the Department of Homeland Security, the mission of the USFA is to reduce life and economic losses due to fire and related emergencies, through leadership, advocacy, coordination, and support. We serve the Nation independently, in coordination with other Federal agencies, and in partnership with fire protection and emergency service communities. With a commitment to excellence, we provide public education, training, technology, and data initiatives.

TABLE OF CONTENTS

Hospital Fire Kills Four Patients
Southside Regional Medical Center
Petersburg, Virginia
December 31, 1994

Local Contacts: Fire Investigator/Inspector Rufus L. Atkins
Fire Investigator/Inspector David M. Jolly
Petersburg Bureau of Fire
400 East Washington Street
Petersburg, Virginia 23803
(804)733-2409

OVERVIEW

On December 31, 1994, a New Year's Eve fire at the Southside Regional Medical Center (SRMC) in Petersburg, Virginia, killed four patients and injured three firefighters and several nurses.[1] The fire was the worst in terms of number of lives lost in a single incident in Petersburg in recent memory.

The fire originated in a patient room on the fourth floor of the hospital shortly after 9 p.m. Local investigators believe that smoking materials were involved in the ignition and that the fire resulted from the patient's actions. They could not determine if the actions which caused the fire were accidental or intentional. The fire grew rapidly due partly to the fuel load in the fire room. A nurse discovered the fire but was unable to extinguish it.

The fire was contained to the room of origin with slight extension into the adjacent corridor. There was heavy smoke and heat damage to the corridor, and heavy smoke infiltrated adjacent patient rooms (see Appendix A for floor diagram). Hospital staff and nurses rescued several patients before firefighters arrived. Three patients in rooms adjacent to the fire room died from smoke inhalation, and one patient in the fire room died from a combination of smoke inhalation and burns. Three firefighters were treated for minor smoke inhalation after their breathing apparatus ran out of air during rescue operations. Several nurses and patients also suffered minor injuries from smoke inhalation.

The section of the hospital that was involved in this incident had smoke detectors in the corridors and manually activated fire alarm stations. Other parts of the hospital complex were protected by automatic sprinklers. Firefighters extinguished the fire with one 1-3/4-inch handline stretched from a standpipe.

---

[1]Two other patients, who were on the fire floor but not in rooms on the corridor of origin, died within several weeks after the fire. At the time of this report, it was unknown whether their deaths were from natural causes or resulted from exposure to smoke.

1

SUMMARY OF KEY ISSUES

Issue	Comments
Fire Cause	Patient smoking in bed was the most likely cause of the fire. Patient failed to comply with hospital no smoking policy. The hospital staff was unable to search the patient's belongings after smoking materials were discovered in the room due to patient privacy rights.
Compartmentation Effectiveness	Failed because door to room of origin was left open. Smoke and heat spread into the corridor.
Sprinkler System	None in area involved in this incident.
Fire Load	Heavy fire load in patient rooms, including highly combustible foam/plastic patient bedding.
Smoke Spread	Rapid spread of heavy smoke into the corridor through the open door. Smoke spread to adjacent rooms through the plenum space and through cracks between doors and door frames, even though the doors to patient rooms were closed.
Evacuation of Patients	Many patients were unable to evacuate without assistance.
Alarm System Performance	Manual pull station activated promptly after discovery.
Construction Code	Appears to comply with code that was in effect when built and fire resistive construction appears to have been maintained fairly well. Some fire resistive features failed early in fire.
Leakage of 100% Oxygen into Room of Origin	Appears to have contributed to the rate of combustion in the room of origin.
Smoke Detection System	None in patient rooms; only in corridor.
Smoke Control Doors	Released to close by alarm system. Contained heavy smoke to one corridor.
Structural Damage	Columns and major beams undamaged. Some light beams and metal decking slightly warped.
Response by Hospital Staff	Prompt and efficient in most regards. Failure to close door to room of origin was a significant factor.
Response by Fire Department	Prompt response and extinguishment saved lives of some trapped patients.
Hospital Fire Evacuation Plan and Drills	Hospital staff conducted regular drills and safety awareness training

LOCATION

The SRMC is located in Petersburg, Virginia, a city of 40,000 people located approximately 25 miles south of Richmond, Virginia. Petersburg is an historic city in an area that is considered a suburb of Richmond.

The Petersburg Fire Department is a career department with 86 members operating out of four stations. Nineteen firefighters are assigned on each shift to staff four engine companies and one ladder company. A battalion chief supervises each shift. Emergency medical services are provided, primarily by volunteer rescue squads; however, the fire department provides first responder medical care and operates one ambulance. The department responds to approximately 3,200 calls per year. It has established mutual aid agreements with several neighboring jurisdictions.

HOSPITAL DESCRIPTION AND FIRE CONTROL SYSTEMS

The original parts of the hospital were constructed in the early 1950s. Since then several additions and renovations have taken place, adding a variety of construction types and fire control systems to the facility. The main hospital is a 468 bed, seven story building with three wings (see Appendix A).

The entire hospital is equipped with a combination smoke detection and manual pull station alarm system. In some areas, smoke detectors are installed in patient rooms; other areas have smoke detectors only in the corridors. The emergency department, which was recently renovated, is protected by an automatic sprinkler system. A Class 3 standpipe system serves all floors; outlets for fire department hoselines are located in hose cabinets in the corridors.

The seven story south wing of the hospital was renovated in 1974. It is a T-shape with three corridors extending out from a nurses' station and elevator lobby. Smoke control doors were installed at the opening of each corridor to provide a separation from the lobby. The doors, which are normally open, are automatically released to close upon activation of the building fire alarm system. An enclosed exit stairway is located at the end of each corridor. On the fourth floor, the T-shaped area referred to as 4- South includes 27 patient rooms. The room where the fire originated is on the south corridor of the 4-South wing, which contains rooms 411 through 421. Each of the patient rooms in this area is equipped for two patients; however, at the time of the fire most of the rooms were occupied by only one patient each.

Construction Details--When the south wing was renovated in 1974, it was required to meet the BOCA building code.[2] The construction of the south wing is classified as "protected non-combustible." The floor/ceiling system incorporates a concrete floor slab supported by steel bar joists. The steel decking and other structural steel members are protected by a suspended ceiling. The suspended ceiling is part of the rated fire resistive assembly, and clips were required to keep the ceiling tiles in place.

The interior partitions are gypsum wall board on metal studs. Above the suspended ceiling, there was a 3 foot plenum space[3] containing wiring, ducts, and oxygen pipes. The plenum space is common above the patient rooms on each side of the corridor. The walls that separate the rooms from the corridor extend through the plenum to the underside of the metal decking of the floor above. The patient rooms could be considered approximately equivalent to a one hour fire-rated compartment, if the room door is closed and the ceiling is in place. Each patient room also had an exterior plateglass window.

There were dampers activated by heat and smoke detectors in the air handling ducts in the plenum space. The air handling system supplied fresh air to the corridors and each room but there was no return air system in the south wing. Heat activated dampers were installed at the point where each duct penetrated the corridor walls above the ceiling to supply air to a single outlet in each room. Air was removed continuously by fans located in each patient bathroom.

The south wing is served by three elevators located near the nurses' station. There were three additional elevators on the north wing of the hospital.

---

[2]At the time of the 1974 renovation of 4-South, the wing was constructed to the BOCA building code, which makes reference to certain NFPA fire codes including NFPA 99, *Standard for Health Care Facilities*. The BOCA code does not reference NFPA 101, *Life Safety Code*.

[3]The plenum space is the area between the suspended ceiling and the next floor.

Fire Alarm System Details--There were two smoke detectors in each corridor of 4-South, but no detectors in the patient rooms. There were two manual pull alarm stations on the south corridor. Activation of the fire alarm system released the smoke control doors and sounded an alarm throughout the hospital. The doors to individual patient rooms did not have self-closing devices and did not close automatically when an alarm was sounded. The 4-South wing was not equipped with an automatic sprinkler system. A hose cabinet supplied by a Class 3 standpipe riser was located in each corridor.

The hospital's fire control room is located in the security office on the ground floor. All alarms are monitored from this location. Four separate fire alarm systems were installed as different parts of the hospital were added or renovated. The fire alarm system had a manual switch with the options of direct connection to a central alarm monitoring service or local monitoring only, but it was not connected to a third party central monitoring station at the time of the fire. The hospital had decided to monitor the alarm locally due to a false alarm problem. In lieu of a fire department response, the normal practice was to have a security guard respond to the activation point to determine if there was an actual emergency before calling the fire department. In this case, a nurse from 4- South notified the hospital switchboard operator via telephone that there was an actual fire, and the switchboard operator notified the fire department. The operator also announced over a hospital wide public address system that "Plan F," the plan for a working fire, was in effect. The activated alarm sounded throughout the entire hospital.

Fuel Load--The fuel load in the room of origin, which measured 10 feet by 12 feet, was typical for the rooms in that section of the hospital. It included:

- Two patient beds

- One dresser and one bedside food table made of laminated fiberboard

- Two small wooden closets affixed to the wall

- Two reclining chairs covered with foam plastic material

- Two divider curtains made of a synthetic fiber

- One plastic wastebasket

- A 19-inch television

- Window curtains

The two patient beds were relatively newly designed "air beds." Each bed incorporated an inflatable mattress and sophisticated control system. The mattress is constructed with inflatable air chambers that can be inflated or deflated to adjust to the needs and comfort of each patient. An electrically operated compressor and pressure controls were mounted in the frame of the bed. Each inflatable mattress was believed to contain between 15 to 20 pounds of foam plastic materials.

The fire room also contained some additional objects such as a patient's suitcase and clothing.

At some point during the fire, the oxygen regulator on the wall in the room of origin melted and may have released a flow of 100 percent oxygen into the room for a short period until it was shut off. After the fire the damaged regulator was tested and was determined to release oxygen at an undetermined rate. The oxygen flow was stopped when a maintenance worker closed the central oxygen valve approximately three to five minutes after the alarm sounded, but before the fire department arrived on the fourth floor.

CAUSE OF FIRE

Investigators determined that the fire originated in Room 418. The patient in Room 418 had been given sedatives to reduce her level of agitation. Nurses had restrained the patient in bed because she had wandered out of the room on at least two occasions that day and was being disruptive to other patients on the floor. The patient broke out of the restraints several times.

The origin of the fire was determined to be the upper half of the patient's bed. After a thorough investigation, the investigators ruled out other possible causes and determined the cause to be the patient smoking in bed or attempting to use matches or a lighter. They were unable to conclusively determine if the fire was caused by a cigarette or a match. The sudden onset of smoke and fire suggests an open flame ignition as opposed to a smoldering ignition which is more often associated with a cigarette. No smoke or fire was noticed when nurses were restraining the patient approximately 10 minutes before the fire was detected. At the time of ignition, the patient may have broken free of the restraints or had enough slack to reach her smoking materials.

DISCOVERY OF THE FIRE AND NOTIFICATION OF THE FIRE DEPARTMENT

The fire at Southside Regional Medical Center occurred on New Year's Eve 1994, which was a Saturday night. At that time, there were six nurses assigned to 4-South. All of the 11 patient rooms on the south corridor were occupied, with one patient assigned to all but two of the rooms. Several relatives were visiting the patient in Room 417. The patients on 4-South were categorized as general nursing care and ranged in age from the mid-40s to late 70s. The age and medical problems of certain patients prevented them from being able to evacuate without assistance; several patients on the south corridor were able to evacuate on their own.

A nurse discovered the fire shortly after 9 p.m. She was alerted by screams from the patient in the room. The nurse opened the door to the room and found the upper half of the bed and the patient on fire. Following emergency procedures, the nurse immediately activated a manual pull station on the south corridor which sounded an audible alarm throughout the hospital.

The fire was reported to the Petersburg 9-1-1 Communications Center by several different sources almost simultaneously. The first 9-1-1 call came from a patient on 4-South at 21:11:30 (see Appendix B for the complete time log of the incident). Seconds later, at 21:11:36, an attendant[4] from a Chesterfield County ambulance which was at the hospital called the Petersburg Communications Center to report that there was fire showing out of a fourth floor window. At 21:11:45, a 9-1-1 call was received from the hospital switchboard.

At 21:12:45, a Petersburg police officer, who was in the hospital when the alarm sounded, arrived on 4-South and radioed to the Communications Center that there was fire.

It is unknown exactly when the fire started. Due to the variances in clocks at the different reporting points and the lack of any recorded times from the hospital's alarm system, an exact timelog of the fire is difficult to reconstruct. The timelog in Appendix B was developed by local investigators. When the fire was reported by the Chesterfield County ambulance attendant at 21:11:36, the fire was already at a free burning phase visible through a fourth floor window.

---

[4]The Chesterfield County ambulance attendant was an off-duty Petersburg firefighter.

INITIAL ACTIONS BY HOSPITAL PERSONNEL

The nurse who discovered the fire was alerted by screams from the patient in Room 418. The nurse opened the door to the room and saw fire involving the upper portion of the bed and the patient. She exited the room, yelled "Fire," and then activated a manual pull station on the south corridor, which sounded an alarm throughout the entire hospital. The nurse re-entered the room and attempted to smother the fire with a blanket but was forced outside due to smoke conditions. Knowing that the patient in Room 418 was placed in a four point restraint system, she re-entered the room to attempt to free the patient by cutting the straps, but she was only able to cut one strap before being forced out of the room by the smoke.

Some of the other nurses began closing doors to patient rooms and evacuating some of the patients from rooms near the room of origin. Several nurses took fire extinguishers to the room of origin, but the rapidly increasing volume of fire prevented them from entering the room and using the extinguishers.

Several hospital personnel responded to the report of a fire on 4-South according to the hospital's emergency response plan. Immediately after the alarm was activated, one nurse telephoned the hospital switchboard to report the fire. The switchboard operator was then responsible for contacting the fire department and announcing the "Plan F" (the actual fire plan) over the public address system, which activated emergency procedures throughout the hospital.

Hospital personnel did not close the door to the room of origin which allowed smoke to fill the corridor very quickly. The nurses managed to either evacuate or close the door to each of the other rooms that opened to the south corridor. Within a few minutes after discovery of the fire, smoke conditions were so bad that the last nurse leaving the corridor had to crawl out. One nurse attempted valiantly to drag a patient out of Room 421, but she was forced to leave the patient in the room and close the door because of the rapidly deteriorating conditions in the corridor. Several patients were evacuated by nurses to stairwell number one, but the rapidly deteriorating conditions in the corridor prevented them from being able to evacuate all of the patients.

A guard and a maintenance worker who responded to the fourth floor after hearing the alarm encountered moderate smoke in the main corridor and very heavy smoke in the 4-South elevator lobby. They attempted to enter the elevator lobby area several times but were forced to retreat behind the protection of smoke control doors. The maintenance worker reached the doors separating the south corridor from the nurses station but was forced to retreat. He was able to shut off the central oxygen valve in the elevator lobby area. This worker estimated that he arrived on 4-South three minutes after the alarm was activated. He reported that the smoke conditions prevented him from seeing the end of the corridor (approximately 75 feet away).

A doctor also responded to the floor and assisted the patients who were being evacuated into the stairwell. He and nurses not directly involved in the initial evacuation assisted later with treatment of patients who were rescued by firefighters.

INITIAL FIRE DEPARTMENT RESPONSE AND FIRST ALARM ACTIONS

The Petersburg Communications Center initially dispatched a full first alarm structural assignment which included Engine 2, Engine 4, Truck 1, an ambulance, and a battalion chief at 21:12 (see Appendix C for alarm assignments). The Chesterfield County ambulance radioed the responding Petersburg Battalion Chief that this was a working fire with flames coming from a window. The battalion chief requested a second alarm before any of the first alarm units arrived. The second alarm response included Engine 5 and Engine 3, the last two Petersburg units, plus a ladder truck on mutual aid from Chesterfield County. Three battalion chiefs and one district chief from Chesterfield County also responded. (Additional units were requested on special alarms later. Refer to Appendix C for a complete list of units that responded to the fire.)

The first arriving Petersburg units went to their assigned positions outside of the hospital, based on the pre-fire plan. Engine 2, the first due company, arrived at 21:15:33 at the main entrance on the west side. All of the first alarm units arrived within one minute of each other.

The objective of Engine 2 was to locate the fire and to initiate interior rescues and fire attack. Arriving at the main entrance on the west side, the crew could not see the fire coming from the fire room, which was on the east side. The building alarm system was sounding when two firefighters from Engine 2 entered the hospital.

The second arriving company, Engine 4 went to the east side, connected to a hydrant, and supplied water to one of the hospital's three standpipe connections.

Truck 1 positioned on the south side of the hospital. Two firefighters from Truck 1 entered the hospital to assist Engine 2. Two other firefighters from Truck 1 set up a ground ladder to rescue the patient from Room 416 who was leaning out of the window. Due to the design of the building, the truck could not be positioned to rescue the individual with its aerial ladder. The rescue was made using a 28 foot ladder which was raised from the roof of a one-story section of the hospital, below the window.

Unit 530, Petersburg Battalion Chief Steve Bowling, established a command post on the east side of the hospital at 21:16:05. For the initial minutes, the Incident Commander's priority was to ascertain the magnitude of the fire, determine what rescues would be necessary, and find out if the fire was extending to other parts of the fourth floor.

Engine 2 reported that the alarm panel indicated an activation of a manual pull station on 4-South. Engine 2's crew then ascended stairwell number one (see Appendix A) to the fourth floor. When the crew reached the fourth floor, they were met by patients and hospital personnel who had been evacuated to the stairwell. They entered the main corridor on the fourth floor and noticed moderate smoke conditions. The crew put on their SCBAs and opened the first set of smoke control doors. Just beyond the doors, they connected a 1-3/4-inch attack line into the Class 3 standpipe connection in the corridor[5]. The firefighters then crawled to the next set of smoke control doors and opened them to find heavy, black smoke, zero visibility conditions, and intense heat. At 21:18:40, Engine 2 radioed to the Incident Commander that they were in the south corridor looking for the fire.

As Engine 2 proceeded down the corridor looking for the fire, they searched rooms and rescued patients. They rescued at least one patient before attacking the fire. When they reached Room 418, the room was fully involved with fire spreading out into the corridor. Engine 2 applied water to the fire at 21:24,[6] 12 minutes after dispatch of the call, and only nine minutes after their arrival on the scene. They were able to gain control of the fire very quickly with one attack line.

---

[5]The 1-3/4-inch attack line was attached to a 50 foot, 2-1/2-inch leader line with a gated wye.

[6] 21:24 is the time that the Incident Commander reported via radio that he could see water being applied to the fire from his vantage point outside the hospital.

Engine 2 and two firefighters from Truck 1 then completed a primary search of the rooms on 4-South. They removed the patients from rooms 417, 419, and 421, all of whom died from smoke inhalation. Several other patients who survived were also evacuated by firefighters. The evacuated patients were carried from their rooms to the stairwell and to unaffected parts of the fourth floor where they were triaged and treated.

The firefighters were using 30 minute SCBA units, and in the course of rescue operations, some of them ran out of air, but many continued to perform rescues because of their concern that the patients must be removed and treated rapidly. Three firefighters suffered smoke inhalation in the rescue process.

Crews from the second alarm units were assigned to check for fire extension on the fifth floor and to remove smoke from the corridors. Rapid smoke removal was important because the longer the smoke remained in the corridors; the more it was sucked through openings around the closed doors to occupied patient rooms.

There was no vertical extension of the fire above the fourth floor; however, the radiant heat on the fifth floor was very noticeable. The concrete floor slab above the fire area remained very hot for some time after the fire. There was a moderate amount of smoke on the fifth floor, with heaviest concentrations in the 5-South area directly above the fire. The second alarm crews also evacuated patients from 5-South and other areas of the hospital affected by smoke conditions. They were assisted by hospital staff.

SMOKE AND HEAT SPREAD

Although the fire was contained to the room of origin with minor extension into the corridor, heavy smoke and heat conditions made the 4- South corridor and the lobby area around the nurses' station untenable within several minutes after discovery of the fire.

The extent of the smoke and heat conditions was evident after the fire. Heavy smoke stains were evident to within two feet of the floor on the south corridor. The smoke detectors and other plastic fixtures in the corridor melted. The smoke control doors separating the south corridor from the nurses' station protected the rest of the floor from major smoke and heat damage. Witnesses reported that there was heavy smoke in the area of the nurses' station and elevator lobby during the incident, possibly from the opening of the smoke control doors as patients were evacuated from the south corridor.

Smoke and heat damage to patient rooms on the south corridor varied significantly. Rooms 411 and 421 suffered heavy smoke damage, indicated by major soot stains on the walls and windows, because the doors to these rooms were open during the fire. The door to Room 411 was not closed by nurses because they had rescued the patient from this room and knew it was empty. The door to Room 421 was closed by nurses, but the patient apparently opened it while trying to escape and collapsed in the doorway.

There was some evidence of smoke leakage between the doors and door frames from the corridor into all of the patient rooms. The rooms on the same side of the corridor as the fire room, i.e., Rooms 416 to 421, appear to have been much more heavily charged with smoke during the fire, while the rooms on the opposite side of the corridor appear to have suffered less smoke damage.

There was evidence of significant smoke spread into Rooms 416 to 421 via the common plenum space above the rooms. This was indicated by soot streaks on many of the ceiling panels, indicating that heavy smoke was leaking into the rooms via gaps between the ceiling panels and the suspension system. The ceiling above the fire failed, allowing smoke and heat to enter the plenum area. The smoke then leaked down through the ceilings, under pressure, into the adjacent rooms. The heat damaged some of the lightweight steel members in the floor assembly immediately above Room 418.

The wall that separates these rooms from the corridor also separates the plenum areas, which limited smoke spread via the plenum to the rooms across the corridor. There was no evidence of smoke spread into any of the rooms via the HVAC system.

LOCATION OF VICTIMS

The patients who died were located in rooms 417, 418, 419, and 421. The patient in Room 418 was burned severely, and skeletal remains were found along with the metal parts of the bed. The patients in rooms 417 and 419, immediately adjacent to the fire room, were found unconscious in their beds, and probably succumbed to smoke inhalation (the Medical Examiner's reports on causes of death were not available at the time of this report). The patient from Room 421 was found unconscious in the partly open doorway to his room.

The patient in Room 416 was awakened by the odor of the smoke and was unable to escape through the corridor. He reported that the smoke in his room was very heavy. He leaned out of his window for fresh air and was rescued by firefighters using a ground ladder.

DAMAGE ASSESSMENT

Room 418 and all of its contents were completely consumed. Only the metal frames of the beds and chairs remained. The suspended ceiling in the fire room collapsed during the fire. The unprotected steel trusses and the floor decking immediately above the suspended ceiling sagged slightly. The top half of the door to Room 418 was burned away. With the exception of a small area near the doorway, all of the floor tiles in Room 418 were consumed.

The south corridor on the 4-South wing suffered extensive smoke and heat damage. The patient rooms adjacent to the south corridor suffered moderate heat and smoke damage. The remainder of the fourth floor beyond the smoke control doors suffered only minor smoke damage. The fifth floor south wing also suffered minor smoke damage. The fire caused approximately $500,000 in damages.

HOSPITAL FIRE DRILLS

The hospital has a comprehensive fire and safety drill program. Each shift drills twice each quarter on emergency procedures. In some cases the shift personnel are drill participants, while in other cases they are observers.

The hospital teaches the standard four-step RACE procedure for handling a fire. RACE stands for:

Remove victims in the room of origin, or remove the fire to the outside (such as in the case of a burning trashcan).

Activate the fire alarm system.

Contain the fire by closing the door to the room of origin and to other rooms and areas.

Evacuate patients.

The 4-South nursing staff on duty the night of the fire had participated in a fire drill two nights before the fire occurred. In general, the staff appeared to have a good understanding of emergency procedures.

The hospital staff response to this fire appears to have been very good in most respects. The nurse who discovered the fire followed most of the emergency action steps that she had been taught in drills. When she discovered the fire, she first activated the alarm system, and then attempted to smother the fire with a blanket and to cut the patient loose from the restraints. Other nurses responded with fire extinguishers but were unable to use them due to the size of the fire and amount of smoke they encountered when they reached the room. The nurses then began evacuating the mobile patients from 4-South and attempted to remove sleeping patients. They closed all of the doors to occupied patient rooms in the south corridor and the two intersecting corridors.

ANALYSIS

This section of the report identifies and analyzes the significant findings that can be drawn from this incident.

Fire Growth and Smoke Spread

Based on accounts of the fire from hospital personnel and physical evidence, it appears that the fire grew rapidly and that flashover in the room occurred within a maximum of five to seven minutes after the fire was first detected by a nurse. Reports from the Chesterfield ambulance crew indicate that the room's window blew out at approximately 21:12. The fire generated large quantities of heavy black smoke which rapidly filled the adjacent area.

The rapid fire growth and smoke spread can be attributed to several factors listed below.

- The beds provided the major source of fuel for the fire. The plastic parts of the bed, particularly the foam plastic in the mattresses provided the fuel for rapid fire growth and heat release, with large volumes of black smoke. The cushions on chairs in the room also contributed to the rapid burning and smoke production.

- The door to the fire room was left open, which allowed the smoke and heat to enter the corridor. This exposed all of the other rooms to smoke and hindered rescue attempts by the staff. Had the door been closed, the bulk of the smoke and heat would have been contained to the room. The fire growth may also have been slowed by the reduction in ventilation.

- The suspended ceiling in the room of origin collapsed when the metal frame structure began to warp, probably around the time of flashover. This provided an opening by which smoke could travel through the common plenum space above Rooms 416-421 and into the rooms.

- An oxygen regulator on the wall melted, releasing 100 percent oxygen into the room at an undetermined rate. The patient was taking oxygen therapy but was not on oxygen at the time of the fire.

    It is believed that the damage to the oxygen regulator occurred after flashover of the patient room. The emergency oxygen shut-off valve at the nurses' station was closed approximately three to four minutes after the fire was discovered. The oxygen may have intensified the fire for a short time, however, it does not appear to have contributed to the rapid fire growth and smoke production prior to flashover.

- The sliding window in the fire room was open approximately six inches at the time of the fire. This provided additional air supply to the fire and may have contributed to a draft through the room and through the open door into the corridor.

Hospital Staff Response

The rapid fire involvement and smoke production appear to have forced the nurses to retreat before they were able to close the door to the fire area. A maintenance worker responding to investigate the alarm closed the emergency oxygen valve for 4-South. Although it would have been preferable for this valve to have been closed immediately upon discovery of the fire, the staff was very busy evacuating patients and closing doors during this period. A general announcement was made over the hospital public address system indicating that there was an actual fire in the building which notified staff throughout the hospital to activate emergency procedures. Hospital personnel, including a doctor, responded to the floor to triage and treat patients removed by firefighters. Patient charts were removed from 4-South by nurses so that other hospital staff would know how to treat patients who had been removed from rooms and disconnected from oxygen and intravenous treatments.

In general, it appears that hospital staff was very familiar with emergency procedures for a fire and, in most cases, followed these procedures. The efforts by the hospital's safety department to train staff about emergency procedures appears to have played a significant role in an efficient handling of this emergency. Unfortunately, in the rush to evacuate 4-South and activate other emergency procedures following the discovery of the fire, the door to Room 418 was left open. This illustrates why it is important to have alarm activated self-closing doors throughout the hospital, including patient room doors.[7]

Based on reports about the size of the fire when it was discovered, it is very likely that it could have been extinguished with one of the water extinguishers kept on the south corridor.

Fire Department Response

Incident Command--Battalion Chief Steve Bowling assumed command on the east side of the hospital immediately after arriving on the scene. Chief Bowling's initial plan was to have the first arriving units execute their planned assignments and to ascertain as quickly as possible the amount of fire and the extent of evacuations and rescues that would be necessary.

Planned assignments, which included the first arriving engine checking the fire annunciator panel and the second arriving engine establishing the water supply and charging the standpipe system, were executed promptly. All of the 14 personnel on the first alarm were busy either fighting the fire or performing rescues for the first 20 minutes. Until other personnel arrived, the Incident Commander did not have the necessary personnel to assign sector officers.

---

[7] NFPA 101 recognizes the value of self-closing doors to patient rooms, but recommends that rooms equipped with self-closing doors also be equipped with an automatic detection system connected to an annunciator panel that indicates which detectors are in alarm. Self-closing doors may make it difficult for hospital personnel to determine the origin of smoke, which is why the automatic detection system is recommended when self-closing doors are used.

A significant fire in an occupied institutional facility is more than one commander can manage alone. Chief Bowling noted that the firefighters from the first alarm companies were too busy fighting the fire and rescuing victims to be able to provide a full evaluation of the conditions inside the hospital. Off-duty command officers were immediately called to respond, but most of them did not arrive on the scene until the fire had been extinguished.

Coordination Between Incident Command and Hospital Staff--The Incident Commander did not have a representative from the hospital at the command post to serve as a liaison until over one-half hour into the incident. As a result, he was unable to resolve some key questions about ventilation systems, elevator problems, and the hospital's immediate needs.

Two specific problems arose that required cooperation and coordination between the fire department and the hospital. First, fire department personnel could not control the elevators initially because they did not have the necessary keys. While firefighters needed elevators to shuttle equipment to the fire floor, hospital staff needed the elevators to evacuate patients. This is an important issue that should be addressed in an evacuation plan.

The second problem dealt with transfer of patients. Several patients from the fire floor suffered injuries that required them to be transferred to other hospitals with burn units. Patients from the fourth and fifth floors were moved to areas of the hospital unaffected by the fire. The hospital had to borrow portable oxygen cylinders from every fire department unit on the scene to treat patients who had been exposed to the smoke.

Hospital Construction

The hospital was built to the 1974 BOCA building code. The codes required institutional buildings to be divided into fire resistive compartments to contain fire spread. The original building was not sprinklered.

The effectiveness of many compartmentation systems tends to decrease over time as various systems are installed or repaired. The effectiveness of the compartmentation on the south wing of SRMC appears to have been largely maintained over more than 20 years.

Each patient room was designed to be a fire resistive compartment capable of containing a fire for approximately one hour. The compartmentation failed early due to the door to the room being left open and the failure of the suspended ceiling early in the fire. The spread of smoke and heat to the corridor and the adjacent rooms resulted in three additional deaths. If the fire and smoke spread had been contained to Room 418 only one death would have occurred.

It is unclear why the ceiling assembly failed during the fire. The ceiling was part of a fire resistive floor ceiling system. One or more of the ceiling tiles, which must be secured in place with pins to function properly, may have not been secured, allowing them to be drawn up and out of the frame of the suspended ceiling by air drafts created by the fire. Due to the amount of damage to the ceiling, however, it was impossible to determine whether each ceiling tile was secured in place with clips.

There were two exits from the corridor, one through the smoke control doors, and the other to the exit stairway.

Hospital Fire Protection Systems

The presence of a sprinkler system in the room of origin would have extinguished or controlled the fire and probably prevented the deaths of three patients in adjacent rooms. Sprinklers were not required under the code when the 4-South corridor was constructed.[8]

On the night of the fire, the fire alarm system was activated by the nurse who discovered the fire and pulled a manual station. A smoke detector in the room of origin would have activated the alarm system more quickly, especially since the door to the room was closed. The detectors located in the corridor offered only limited protection; it is ideal to have a detector in every patient room to provide the earliest possible alert. This is especially important in hospitals and other facilities where patients may be unconscious, sedated, or otherwise unable to detect a fire on their own. The increasing number of electrical appliances in patient rooms is another reason for locating detectors in each room.

The hospital alarm systems used to be monitored by a central station at night, but this practice had recently ceased due to problems with false alarms. Hospital policy is to report an alarm to the switchboard; security officers check the annunciator panel and respond to investigate an alarm. If the switchboard receives a report of an actual fire, however, its policy is to notify the fire department immediately. At the time of the fire, the alarm system was not connected to a central station, and notification depended on a call to the fire department by the switchboard operator. It is preferable to connect the alarm to a central station or to the fire department, as is recommended by NFPA 101.

The alarm system on the south corridor of 4-South includes two smoke detectors and pull stations. When either is activated, a general alarm sounds throughout the hospital. The alarm system does not record an alarm activation time, nor does it differentiate between which alarm was first activated because the detectors and pull stations are on the same circuit. When the fire alarm system activates, the smoke control doors in the corridors close automatically. Patient room doors, however, are not on this system.

HVAC System

The hospital's ventilation system supplies fresh air to the patient rooms via metal ducts that run through the plenum space. Air is removed from each patient bathroom by fans that run constantly. There is no other central ventilation system. Cooling and heating are provided by individual fan-coil units in each room.

The fresh air supply fans were supposed to automatically shut down when the fire alarm activated, but this function was disabled at the time of the fire. A hospital technician manually shut down the fans approximately 30 minutes after the fire was controlled. The failure of this system to shut down may have supplied fresh air to the fire; but this contribution does not appear to have played a significant role in fueling the fire. Nonetheless, the status of the building systems should be readily available to fire department personnel.

---

[8] NFPA 101, *Life Safety Code*®, recommends that existing highrise health care occupancies over 75 feet tall be protected throughout by an automatic sprinkler system. Southside Regional Medical Center is 7 stories tall, making it very close to the minimum height to qualify as a highrise. SRMC was not required to retrofit to meet the Life Safety Code standards.

Smoke stains around the bathroom exhaust fans indicate that the fans drew smoke out of rooms. According to one patient, the fans did not remove the smoke as quickly as smoke filled the rooms.

LESSONS LEARNED

1. **A limited fire protection system in an institution offers only minimal protection.**

    The most important lesson that should be derived from this incident is the importance of installing sprinkler and smoke detection systems throughout institutional facilities where occupant evacuation is difficult and time consuming. The smoke detectors in the corridors did not provide sufficient time for hospital personnel to respond to try to control the incident since the door to the room of origin was closed when the fire started. Even when smoke detection is installed to provide early warning, a real possibility exists that a rapidly growing fire can occur in a patient room with a heavy fire load. In this case, it is preferable to equip every room with a sprinkler system to control the fire.

2. **Preparation of hospital staff for a fire is an integral part of the fire protection.**

    In this incident, the hospital staff was regularly trained and well versed in emergency procedures for a fire. Some of the staff performed heroic rescues that probably saved the lives of some of the patients. Even with good training, the rapid growth of a fire and accumulation of smoke may prevent hospital staff from being able to perform all emergency procedures, including closing all of the doors and evacuating patients. In this case, the door to the room of origin was left open, due to a variety of factors. This reinforces the importance of having automatic extinguishing systems in every room of an institution.

3. **The intent and effectiveness of compartmentation is destroyed when a door or other opening to a room is left open.**

    An automatic, self-closing door system on the doors to patient rooms would have been beneficial in this incident The effectiveness of compartmentation features was destroyed by an open door which allowed smoke and heat spread into the corridor, and which provided a source of fresh air for the fire. The suspended ceiling, which was designed to be fire-rated, failed early in the fire, also destroying the compartmentation effectiveness.

4. **Smoke control doors are an important fire protection feature to control smoke spread.**

    This fire demonstrated that smoke spread can be effectively controlled by smoke control doors and closed doors. Smoke spread via the corridor into other rooms on the corridor was controlled effectively by hospital personnel closing doors to patient rooms. The self-closing smoke control doors at one end of the corridor contained the heavy smoke damage to only one corridor on 4-South wing.

5. **Smoke can spread readily through a plenum space to adjacent areas even though doors are closed.**

    Smoke spread very rapidly, through the common plenum space above six rooms on the corridor due to rapid failure of the ceiling in the room of origin. This allowed smoke to fill the plenum space and to spread into adjacent rooms. Common plenum spaces allow for rapid smoke spread. Suspended ceilings should not be expected to contain smoke due to leakage between tiles and the ceiling frame.

6. **A working institutional occupancy fire places tremendous resource requirements on fire departments.**

   The evacuation of institutional facilities including hospitals, nursing homes, schools, and jails is very laborious and will generally require more firefighting personnel than is needed at fires in other types of occupancies. Initial dispatch assignments for these types of occupancies should include additional resources than normally dispatched for other types of occupancies.

7. **Pre-fire planning facilitates fireground operations.**

   The fire department had developed planned assignments for response to the hospital. Most of the firefighters were familiar with the interior layout, including where standpipe connections were located. Pre-fire planning creates an organized structure of response before the chaos of an emergency incident occurs. It also allows for quick, efficient operations without having delays created by unfamiliar surroundings. This fire required the resources of mutual aid departments. Mutual aid departments should participate in pre-fire planning of institutions or other target hazards outside of their jurisdiction to which they might be summoned.

8. **Command at an institutional occupancy fire requires a high level of coordination and cooperation between the fire department and the institutional staff.**

   The Incident Commander will need the assistance of institutional staff to manage the incident efficiently. Institutional staff, such as a building engineer and medical or administrative personnel, will be necessary to answer questions, to assess needs, and to assist with patient treatment and transfer. The pre-fire plan of an institution should designate a liaison to the Incident Commander to assist with coordination. The liaison should be equipped with the appropriate resources such as keys and a knowledge of the building's mechanical systems. This person should be trained to report to the command post or first arriving unit as part of fire drill procedures.

ADDITIONAL INFORMATION OF INTEREST

As a result of this fire, the Commonwealth of Virginia enacted requirements under State code 3699.9:1 which requires all hospitals in the State to be fully sprinklered by January 1, 1998.

APPENDIX A

Hospital Floor Plan

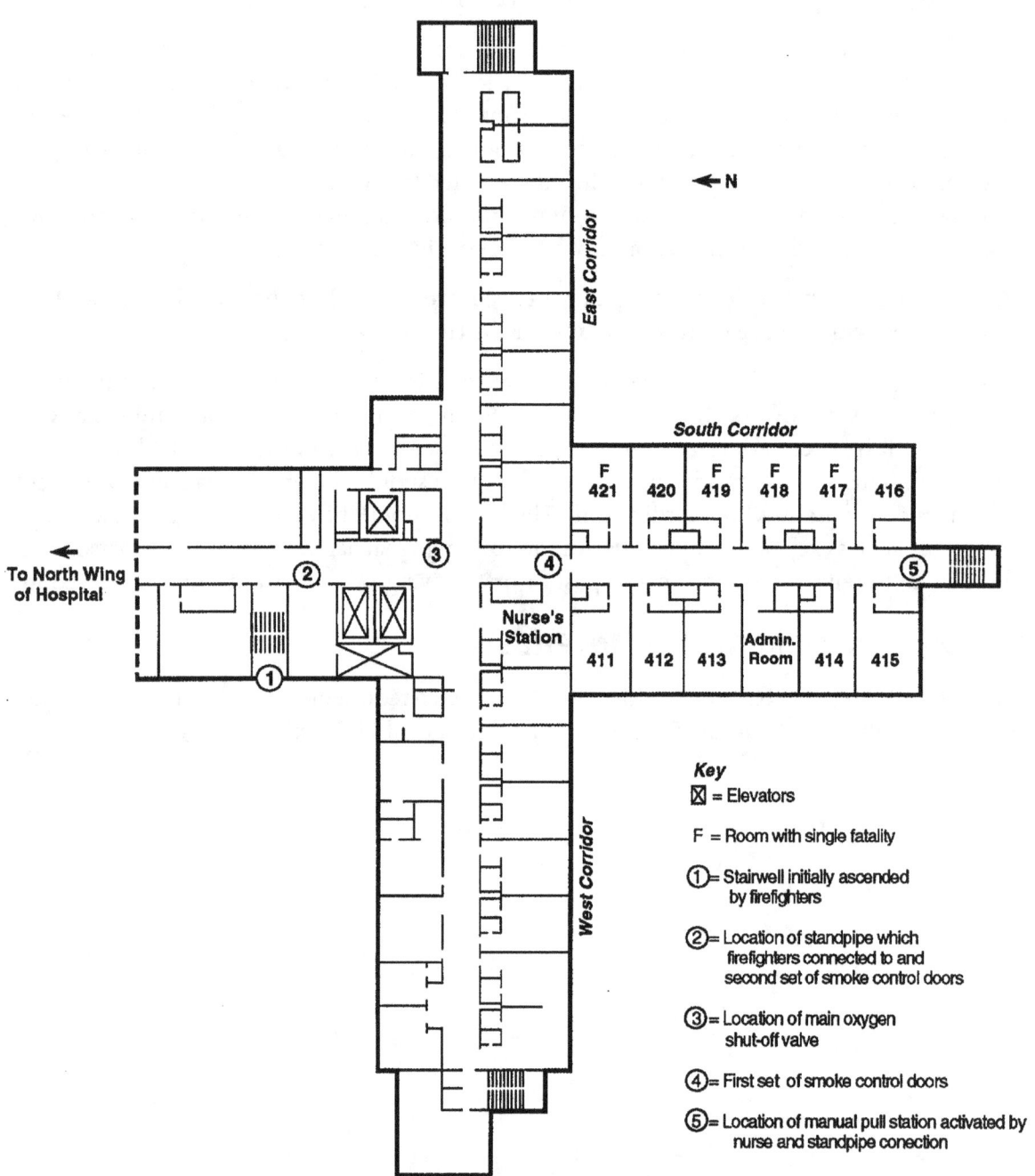

Four-South Floor Plan

APPENDIX B

Time Log of Fire at Southside Regional Medical Center

December 31, 1994

TIME	ACTIVITY
21:11:30	9-1-1 phone call reporting fire received at Petersburg Emergency Communications Center from patient on 4th floor of hospital.
21:11:36	9-1-1 phone call reporting fire received at Petersburg Emergency Communications Center from Chesterfield County ambulance attendant outside of hospital reporting fire.
21:11:45	9-1-1 phone call reporting fire received at Petersburg Emergency Communications Center from hospital switchboard operator.
21:12:20	Dispatch of first alarm.
21:12:45	Petersburg Police Unit 7103 radios Petersburg Emergency Communications Center from 4th floor of hospital reporting fire.
21:13:02	Dispatch of first alarm complete.
21:13:39	First alarm units responding.
21:13:44	Chesterfield Fire Department ambulance radios Petersburg Unit 530 (Battalion Chief) to report working fire on 4th floor with fire evident from outside of hospital.
21:14:00	Second alarm requested by Unit 530.
21:14:14	Petersburg Emergency Communications Center requests response of, off-duty staff officers.
21:14:30	Dispatch of second alarm.
21:15:33	Engine 2 on scene, nothing showing West side.
21:15:54	Fire/Rescue 4 (Petersburg Ambulance) on scene, fire and smoke evident East side. Remainder of first alarm units on scene within next minute.
21:16:05	Unit 530 on scene, establishes Command on East side.
21:17:06	Command advises Engine 2 of person trapped.
21:17:39	Second alarm units responding.
21:18:40	Engine 2 advises it is on 4th floor looking for fire.
21:19:40	Unit 570 (Asst. Chief of Support Services) responding.
21:21:25	Command requests Colonial Heights Fire Department Air Unit (mutual aid).

TIME	ACTIVITY
21:21:38	Unit 507 (Safety Officer) responding.
21:23:48	Chesterfield Fire Unit 127 on scene (75-foot Tower Ladder).
21:24:17	Patient in Room 416 being removed by ladder rescue.
	Command advises he can see line working in room of origin.
21:24:40	Unit 507 on scene.
21:24:54	Unit 570 on scene.
21:27:54	Patient from Room 416 safely on ground.
21:28:05	Air bottles requested on 4th floor.
21:28:45	Colonial Heights Air Unit on scene.
21:31:15	Fire knocked down.
21:33:17	Chesterfield Engine 123 (special request) responding (mutual aid).
21:34:00	Chesterfield Engine 124 (special request) responding (mutual aid).
21:34:20	Petersburg Emergency Communications Center advises Command that Unit 501 is with Unit 520.
21:37:25	Unit 570 requests landline by Unit 501.
21:38:48	Chesterfield Engine 123 on scene.
21:39:33	Chesterfield Engine 124 on scene.
21:41:05	Engine 2 requests oxygen supply to 4th floor be turned off and that 4th floor was all clear.
21:42:21	Signal F (fatality) located in Room 418 (room of origin).
21:42:54	Unit 507 requests notification of hospital safety director.
21:43:10	Red Cross requested.
21:50:15	Command requests update. Truck 1 advises 4th floor all clear, and that Chesterfield Fire Units will perform a secondary search of 4th floor.
21:55:45	Rehab established in hospital cafeteria.
22:07:39	Fire under control.
02:53:00	Last fire unit in service (Investigation Team remained on scene).

APPENDIX C

Alarm Assignments[9]

First Alarm Units

Engine 2, Engine 4
Truck 1
Fire/Rescue 4 (Ambulance)
Unit 530 (Battalion Chief)

Second Alarm Units

Engine 3, Engine 5
Truck 127 (Chesterfield County Tower/Ladder)
Unit 501, 507, 570 (Off-duty Petersburg Chiefs)
4 Chiefs (Chesterfield County)

Special Alarm Units

Engine 123, Engine 124 (Chesterfield County)
Truck (Chesterfield County)
Colonial Heights Fire Dept. Air Unit

---

[9] Bold, italicized units are Petersburg Bureau of Fire units.

APPENDIX D

Photographs of Fire Scene

There was heavy smoke and heat damage around the door to the room of origin.
The top portion of the door burned away in the fire.

Appendix D (continued)

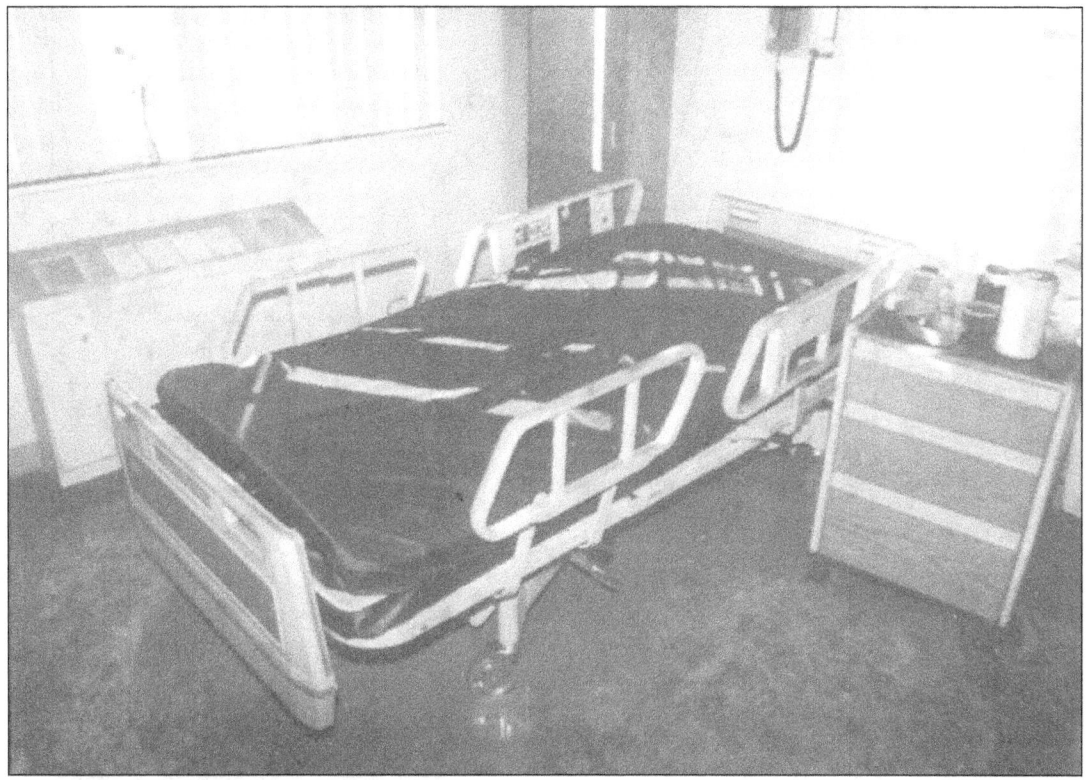

The beds in the hospital contained heavy amounts of plastic and foam parts. Only the metal frame of the patient's bed was recognizable in the room of origin.

Appendix D (continued)

This photo shows the extent of smoke spread in the south corridor, looking down from the room of origin. At the end of the corridor is an exit, a hose cabinet containing a Class 3 standpipe connection and occupant use-hose, and a fire extinguisher, as well as the manual pull station that was activated by a nurse.

Appendix D (continued)

All but a small portion of the floor tile in the room of origin was consumed by the fire. This portion was partially shielded by the door to the room, which was open during the fire. The straight line pattern of burned away tile next to the right side of the doorway indicates the position of the door during the fire.

Appendix D (continued)

Smoke streaks on the ceiling indicate that smoke was entering the rooms adjacent to the fire room through the ceiling suspension system.

Appendix D (continued)

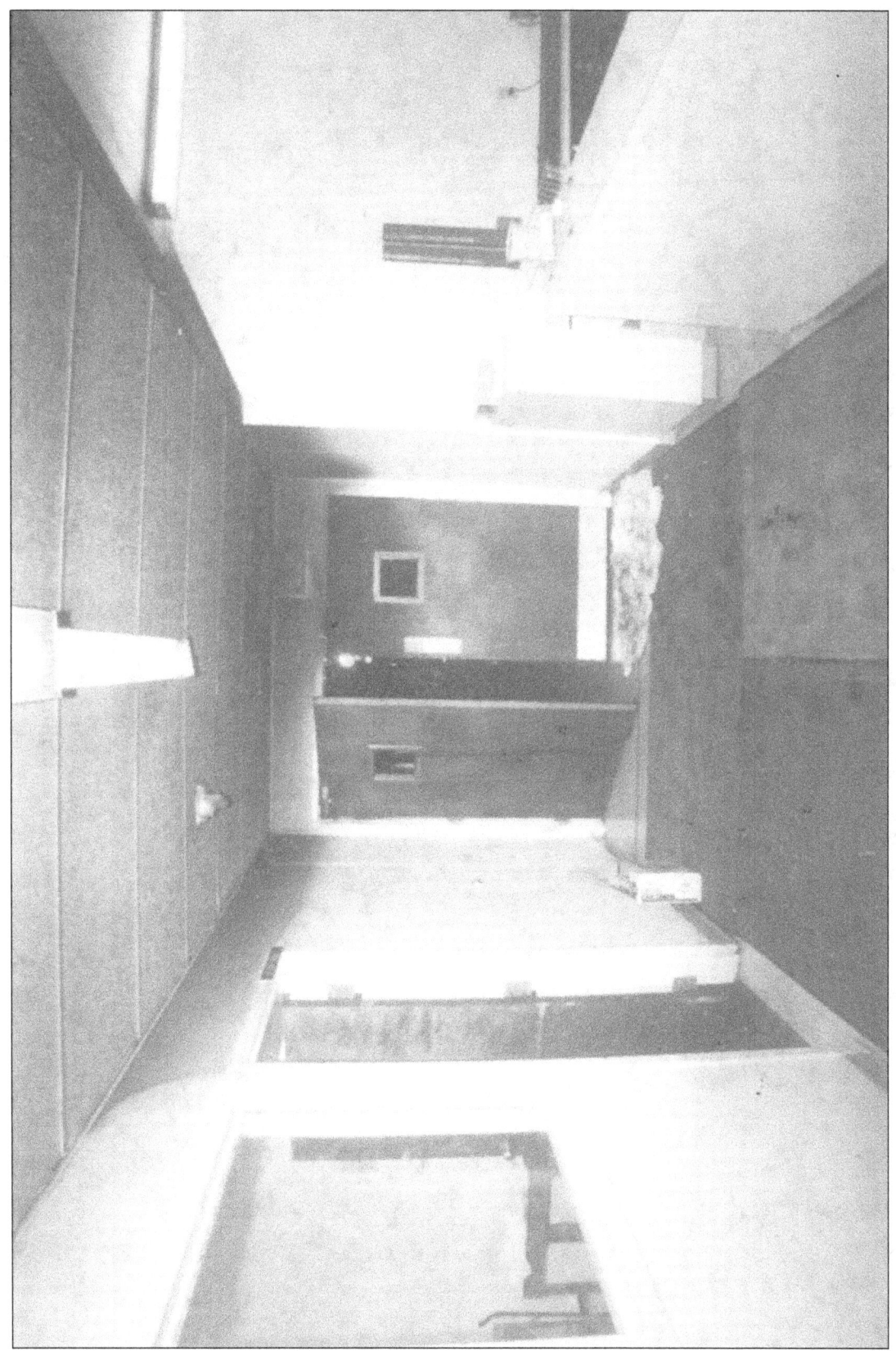

This photo looks down toward the south corridor with the nurses station on the right. The self-closing smoke control doors contained the bulk of the smoke and heat conditions to the corridor of origin, but some smoke escaped into the nurses station area as victims were evacuated through the doors.

Appendix D (continued)

There was very heavy damage to the room of origin. Some structural members were warped by heat after the suspended ceiling collapsed.

Appendix D (continued)

Room 421 which was located on the south corridor suffered heavy smoke damage. Nurses closed the door to this room, however, investigators believe the patient opened the door while attempting to escape. The patient was found lying in the doorway by the firefighters.

Appendix D (continued)

This photo shows the heavy damage to the fire room, Room 418 (on the right), and the adjacent room, Room 419 (on the left). Nurses closed the door to Room 419, but heavy smoke infiltrated the room through the suspended ceiling and cracks around the edges of the door. The patients in both of these rooms died.

Appendix D (continued)

The first floor of the hospital extended out from the upper floors of the building, which blocked access to the fourth floor windows by an aerial ladder. Firefighters used a ground ladder from the roof of the single story wing to rescue a patient from a window located two rooms to the left of the room of origin.